~A BINGO BOOK~

Earth Science Bingo Book

COMPLETE BINGO GAME IN A BOOK

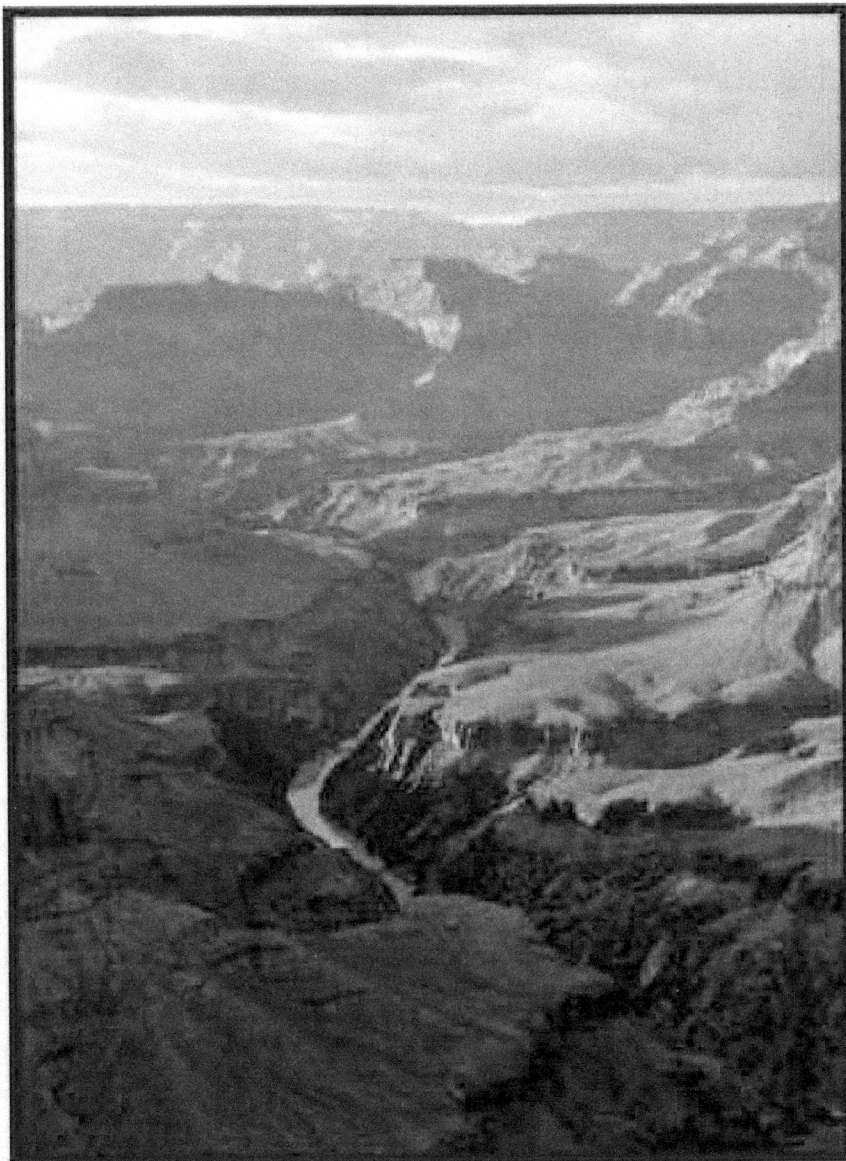

Written By Rebecca Stark

ISBN 978-0-87386-445-9

Educational Books 'n' Bingo

Printed in the U.S.A.

EARTH SCIENCE BINGO
Directions

INCLUDED:

List of Terms

Templates for Additional Terms and Clues

2 Clues per Term

30 Unique Bingo Cards

Markers

1. **Either cut apart the book or make copies of ALL the sheets. You might want to make an extra copy of the clue sheets to use for introduction and review. Keep the sheets in an envelope for easy reuse.**

2. Cut apart the call cards with terms and clues.

3. Pass out one bingo card per student. There are enough for a class of 30.

4. Pass out markers. You may cut apart the markers included in this book or use any other small items of your choice.

5. Decide whether or not you will require the entire card to be filled. Requiring the entire card to be filled provides a better review. However, if you have a short time to fill, you may prefer to have them do the just the border or some other format. Tell the class before you begin what is required.

6. There are 50 topics. Read the list before you begin. If there are any topics that have not been covered in class, you may want to read to the students the topic and clues before you begin.

7. There is a blank space in the middle of each card. You can instruct the students to use it as a free space or you can write in answers to cover topics not included. Of course, in this case you would create your own clues. (Templates provided.)

8. Shuffle the cards and place them in a pile. Two or three clues are provided for each topic. If you plan to play the game with the same group more than once, you might want to choose a different clue for each game. If not, you may choose to use more than one clue.

9. Be sure to keep the cards you have used for the present game in a separate pile. When a student calls, "Bingo," he or she will have to verify that the correct answers are on his or her card AND that the markers were placed in response to the proper questions. Pull out the cards that are on the student's card keeping them in the order they were used in the game. Read each clue as it was given and ask the student to identify the correct answer from his or her card.

10. If the student has the correct answers on the card AND has shown that they were marked in response to the *correct questions,* then that student is the winner and the game is over. If the student does not have the correct answers on the card OR he or she marked the answers in response to *the wrong questions,* then the game continues until there is a proper winner.

11. If you want to play again, reshuffle the cards and begin again.

Have Fun!

TERMS INCLUDED

AQUIFER

ATMOSPHERE

BASALT

BEDROCK

CALDERA

CANYON

CAVERNS

CONTINENT

CONTINENTAL DRIFT

CORE

EARTH SCIENCE

CRUST

EARTHQUAKES

EQUATOR

EROSION

ERUPTIONS

FAULT

FOLDS

FUMAROLES

GEOLOGY

GEYSERS

GLACIER

HEMISPHERE

HYDROSPHERE

IGNEOUS

LATITUDE

LAVA

LITHOSPHERE

MAGMA

MANTLE

METAMORPHIC

MINERALS

FREDERICH MOHS

MOUNTAINS

PLANET

PLATEAUS

PROPERTIES

QUARTZ

RING OF FIRE

ROCKS

SEAFLOOR SPREADING

SEDIMENTARY

SEISMIC WAVES

TECTONIC PLATES

TEPHRA

TIDES

TSUNAMIS

VOLCANOES

WATER CYCLE

WEATHERING

Additional Terms

Choose as many additional terms as you would like and write them in the squares. Repeat each as desired.
Cut out the squares and randomly distribute them to the class.
Instruct the students to place their square on the center space of their card.

Clues for
Additional Terms

Write three clues for each of your additional terms.

_____ 1. 2. 3.	_____ 1. 2. 3.
_____ 1. 2. 3.	_____ 1. 2. 3.
_____ 1. 2. 3.	_____ 1. 2. 3.

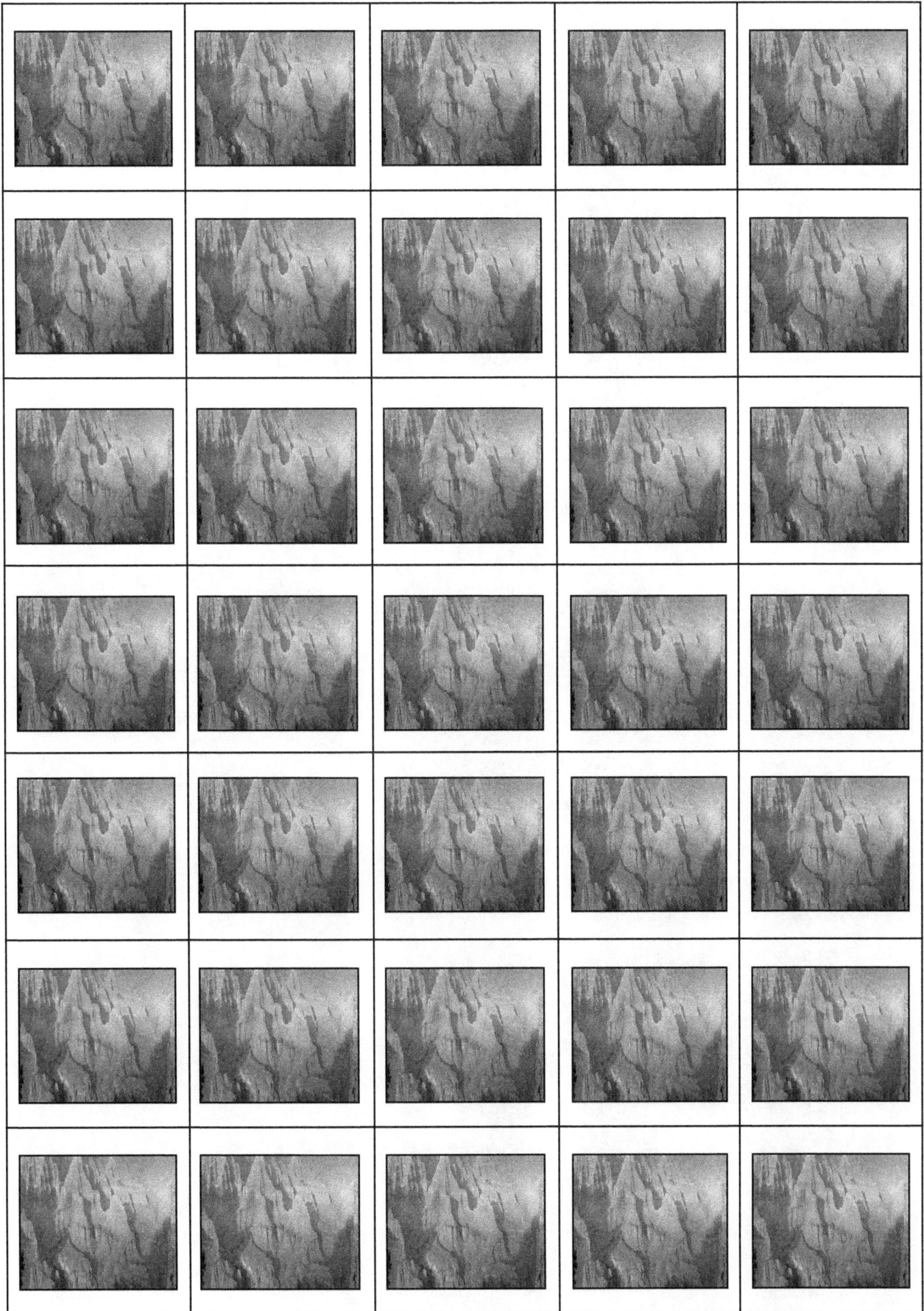

AQUIFER 1. It is an underground bed or layer of Earth, gravel, or porous stone that yields water. 2. The surface of saturated material in one is known as the water table. 3. The vast but shallow underground water table located beneath the Great Plains in the United States is known as the Ogallala ___.	**ATMOSPHERE** 1. This is the whole mass of air surrounding the Earth. 2. It is the gaseous envelope of a celestial body. 3. It is the mass of air surrounding the Earth, composed largely of oxygen and nitrogen.
BASALT 1. It is a hard, dense, dark, volcanic rock. 2. It is the most common form of solidified lava. 3. This solidified lava is a dense, dark-grey, fine-grained igneous rock	**BEDROCK** 1. This is the general term for the solid rock that underlies loose material such as soil, sand, clay, or gravel. 2. It is the solid, unweathered rock beneath the surface deposits of soil. 3. It is the *consolidated* rock beneath the planet's surface as opposed to the area of broken and weathered *unconsolidated* rock above.
CALDERA 1. This is a large crater formed by a volcanic explosion or by the collapse of a volcanic cone. Its diameter is many times that of the vent. 2. Crater Lake in Oregon lies inside one. 3. This term for a basin-shaped volcanic depression comes from the Spanish word for "cauldron."	**CANYON** 1. It is a deep valley between cliffs; it is often carved from the landscape by a river. 2. This landform is much more common in arid areas because weathering has a greater effect in those zones. Their walls are often formed of resistant sandstone or granite. 3. The Grand ___ is a steep-sided gorge carved by the Colorado River.
CAVERNS 1. These large underground chambers belong to a category of caves called solution caves. 2. These large caves are formed by the dissolution of soluble rocks such as limestone, dolomite, gypsum and salt. 3. The ones in Carlsbad, New Mexico, were formed when sulfuric acid dissolved the surrounding limestone.	**CONTINENT** 1. It is one of the principal landmasses, usually regarded as Africa, Antarctica, Asia, Australia, Europe, North America, and South America. 2. The term often includes not only the 7 large masses of land but also the shallow, submerged adjacent areas and the islands on the shelf. 3. It is narrowly defined as a continuous mass of land.
CONTINENTAL DRIFT 1. This refers to the movement of the continents relative to one another. 2. The development of the theory of plate tectonics in the 1960s provided a geological explanation for it. 3. Alfred Wegener is noted for his theory of this. In 1912 he hypothesized that the continents once formed a single landmass, broke up, and slowly drifted to their present locations.	**CORE** 1. The solid inner part is generally believed to be composed primarily of iron and some nickel. 2. The existence of a solid inner part distinct from the liquid outer part was discovered in 1936 by seismologist Inge Lehmann. 3. The inner part rotates in the same direction as the rest of the planet but completes its rotation in about two-thirds of a second less.

Earth Science Bingo

CRATERS

1. They are steep-sided, usually circular, depressions.
2. Calderas are large ones formed by an explosion or collapse at a volcanic vent.
2. Impact ones are the result of the high velocity impact of two celestial bodies, such as the impact of a meteorite upon the surface of a planet.

CRUST

1. It is the outermost solid shell of a planet or moon.
2. Earth's is composed of a variety of igneous, metamorphic, and sedimentary rocks.
3. Earth's thin one is only about 3 to 5 miles thick under the oceans and about 25 miles thick under the continents.

EARTHQUAKES

1. They result from a sudden release of energy in the Earth's crust that creates seismic waves.
2. The point on the Earth's surface that is directly above the focus, or hypocenter, of one is called its epicenter.
3. In 1935 Charles F. Richter devised a scale to describe their intensity.

EQUATOR

1. It is an imaginary line on the Earth's surface equidistant from the North Pole and South Pole.
2. This imaginary line divides the Earth into the Northern Hemisphere and the Southern Hemisphere.
3. It is one of the five main circles of latitude. It is the only line of latitude that is a great circle.

EROSION

1. It is the carrying away, or displacement, of solids, usually by wind, water, ice or waves.
2. This natural process is often increased by poor human land use such as deforestation and overgrazing.
3. Improved land-use practices can limit this process. Terrace-building and the planting of trees are two examples of what can be done.

ERUPTIONS

1. They are the sudden occurrences of a violent discharge of steam and volcanic material.
2. Types vary from tranquil lava emissions to extremely violent, explosive occurrences.
3. Their variability is largely related to the composition of the magma and the amount of water present.

FAULT

1. It is a fracture in the continuity of a rock formation caused by a shifting or dislodging of the Earth's crust.
2. Earthquakes are caused by energy release during rapid slippage along one of these.
3. One that runs along the boundary of a tectonic plate is called a transform ___.

FOLDS

1. They are bends in a layer of rock caused by forces within the Earth's crust. They usually occur in a series and look like waves.
2. Anticlines are those in which the oldest rocks are in their core, or center.
3. Synclines are those in which the youngest rocks are in their core, or center.

FUMAROLES

1. These are small openings from which hot gases and vapors are released into the air.
2. Unlike geysers and hot springs, there is little water available in their vent system. Water that does enter is converted to steam and other gases.
3. Emissions from these small openings include hydrogen sulfide, which causes the "rotten-egg" odor often associated with them.

GEOLOGY

1. It is the study of the composition, structure, physical properties, history, and the processes that shape Earth's components.
2. Scientists who focus on this branch of science are called geologists.
3. Mineralogy, Paleontology. and Petrology are 3 subdisciplines of this branch of science.

Earth Science Bingo

© Barbara M. Peller

GEYSERS

1. They are hot springs characterized by an intermittent turbulent discharge of water. They require a volcanic heat source.
2. A famous one is Old Faithful in Yellowstone National Park in Wyoming.
3. They requires intense heat, which comes from magma; underground water and a reservoir to hold it; and fissures, fractures or other openings.

GLACIER

1. It is a large body of ice that moves slowly down a slope or valley or spreads outward on a land surface in response to gravity .
2. A moraine comprises the rocks and soil carried and deposited by one.
3. An ice cap is a dome-shaped one that forms on an extensive area of relatively level land and flows outward from its center.

HEMISPHERE

1. It is any half of the Earth.
2. The prime meridian divides the Earth into the Eastern and Western ones.
3. The equator divides the Earth into the Northern and Southern ones.

HYDROSPHERE

1. It describes the collective mass of water found on, under, and over the surface of a planet.
2. The ocean is the bulk of Earth's ___.
3. In addition to the oceans, it includes inland seas, lakes, and rivers; rain; underground water; ice, as in glaciers and snow; and atmospheric water vapor, as in clouds.

IGNEOUS

1. These rocks are formed by solidification of cooled magma with or without crystallization.
2. Intrusive ___ rocks form in magma chambers that are deep below the surface. Granite is the most common example.
3. Extrusive ___ rocks form when magma exits and cools outside of or near the Earth's surface. Pumice, obsidian, and basalt are examples.

LATITUDE

1. It is the angular distance north or south of the Earth's equator. Lines of ___ are also called parallels.
2. Because it determines how much sun an area gets, it is one of the most important factors determining climate.
3. The higher the ___, the farther from the equator a place is located.

LAVA

1. It is molten rock that reaches the Earth's surface through a volcano or fissure.
2. While still beneath the surface of the Earth, this molten rock is called magma.
3. Although viscous, it can flow great distances before cooling and solidifying. When it solidifies, it forms igneous rock.

LITHOSPHERE

1. This includes the crust and the uppermost mantle.
2. Just below it is the asthenosphere, the deeper part of the upper mantle.
3. This layer is broken into tectonic plates which move independently relative to one another.

MAGMA

1. This molten rock forms within the upper part of the Earth's mantle and contains liquids, crystals, and dissolved gases
2. Once this molten rock reaches the surface of the Earth—or other terrestrial planet—it is called lava.
3. This molten rock is stored in chambers inside a volcano.

Earth Science Bingo

MANTLE

1. It is a highly viscous layer directly under the crust and above the outer core.
2. The distinction between the crust and this layer is based on chemistry, rock types, flow and seismic characteristics.
3. This layer of Earth, which extends to a depth of about 1,800 miles, accounts for about 85% of the total weight and mass of the Earth.

METAMORPHIC

1. This kind of rock is formed when igneous or sedimentary rock is changed by pressure, heat, or chemical reactions.
2. Marble is one; it starts as limestone or dolomite. Others are gneiss and quartzite.
3. When a rock is exposed to great heat or pressure, it changes mineral composition and texture and becomes a new ___ rock.

MINERALS

1. Rocks are aggregates of these. Some are valued as gems because of their hardness, color and beauty.
2. They are naturally occurring, inorganic solids with their atoms arranged in a definite pattern and with a definite chemical composition.
3. Most are silicates, meaning they combine with oxygen and silicon. Quartz is an example.

FREDERICH MOHS

1. This Austrian geologist devised a scale that characterized the hardness of minerals.
2. In 1812 he suggested that geologists use the same minerals as standards for hardness ratings resulting from scratch tests.
3. When devising his scale, he assigned the softest mineral the number one and the hardest mineral the number 10.

MOUNTAINS

1. These landforms extend above the surrounding terrain in a limited area and have a peak.
2. They are built through the deformation, or "folding," of Earth's crust.
3. When 2 sections of the lithosphere collide causing 1 or both to fold up like an accordion, the crust folds and becomes deformed. These landforms are the result.

PLANET

1. Earth is 1 of the 4 terrestrial ones. All 4 have the same basic structure: a central metallic core, mostly iron; a silicate mantle ; and a crust with canyons, craters, mountains, and volcanoes.
2. This kind of celestial body orbits a star and is massive enough to be rounded by its own gravity.
3. A terrestrial ___ is composed mostly of silicate rocks and is also called a rocky one.

PLATEAUS

1. These large highland areas of fairly level land are separated from surrounding land by steep slopes.
2. Some lie between mountain ranges, and others are higher than the surrounding land.
3. Low ones are often used for farming; high ones are often used for livestock grazing.

PROPERTIES

1. Geologists identify rocks and minerals by their characteristics, or ___.
2. Hardness, streak, fracture, color and luster are some ___ geologists use to identify minerals.
3. The following describes the ___ of gold: It is a soft, yellow, corrosion-resistant element; the most malleable and ductile metal; and a good thermal and electrical conductor.

QUARTZ

1. It is the most abundant mineral in Earth's continental crust. Its chemical formula is SiO_2. The pure variety is colorless or white, but there are many colored varieties.
2. Amethyst is the purple gem variety.
3. Some colored varieties include aventurine, citrine, onyx, agate and jasper.

RING OF FIRE

1. This is an area of frequent earthquakes and volcanic eruptions that encircles the basin of the Pacific Ocean.
2. This area has 452 volcanoes and is the location of about 3/4 of the Earth's active and dormant volcanoes.
3. About 90% of the Earth's's earthquakes occur in this area.

Earth Science Bingo

ROCKS

1. They are naturally occurring aggregates of minerals and/or mineraloids.
2. Petrology, the scientific study of ___, is divided into igneous petrology, sedimentary petrology, metamorphic petrology & experimental petrology.
3. They are classified by their mineral and chemical composition, the texture of their particles, and the processes that formed them.

SEAFLOOR SPREADING
1. This occurs at divergent plate boundaries. Molten material rises along a rift zone and spreads out at the surface building new oceanic crust.
2. The mid-oceanic ridge is the primary site for this process that leads to the creation of new ocean floor.
3. ___ creates the rugged volcanic landscape of a mid-ocean ridge along the plate boundary.

SEDIMENTARY
1. These rocks are formed when layers of eroded earth are pressed down. They form where sand, mud, and other types of sediment collect, such as beaches, rivers, and the ocean.
2. Limestone is this type of rock. It is made from the mineral calcite.
3. Sandstone is ___ rock. It is made from small grains of the minerals quartz and feldspar.

SEISMIC WAVES
1. These waves of energy result from the sudden breaking of rock within the earth or an explosion. They are recorded on seismographs.
2. Primary (P) ___ travel the fastest and are called compressional because of the pushing and pulling.
3. Secondary (S) ___ travel more slowly. They shake the ground perpendicularly to the direction they are moving.

TECTONIC PLATES
1. Earth's lithosphere is divided into 6 huge, rigid ones and several smaller ones.
2. At divergent boundaries 2 of these move apart from one another and new crust is formed. At convergent boundaries they collide and may cause earthquakes, volcanoes or folding.
3. At shear boundaries, 2 slide past one another and crust is neither created nor destroyed.

TEPHRA
1. This is the general term for material produced by a volcanic eruption. At less than 2 millimeters in diameter, volcanic ash is the smallest type.
2. Lapilli, or volcanic cinders, are those between 2 and 64 millimeters in diameter.
3. Volcanic bombs are the largest. They are greater than 64 millimeters in diameter.

TIDES
1. This is the rising and falling of Earth's oceans and other waters caused by gravitational forces of the moon and to a lesser extent the sun.
2. Neap ___ are weak and occur between the first and third quarters of the moon
3. Spring ___ occur near the times of the new moon and full moon and have the most variation in water level.

TSUNAMIS
1. These great sea waves occur when ocean water is displaced by an underwater earthquake, volcanic eruption or other phenomenon.
2. Their small wave height at sea and very long wavelength, cause them to go unnoticed at sea.
3. Those caused by an earthquake in the Indian Ocean in 2004 resulted in the death of more than 225,000 people.

VOLCANOES
1. ___ are vents in Earth's surface through which magma and associated gases and ash erupt; they are also the structures created as a result.
2. The eruption of low-viscosity lavas that flow far from the vent result in shield ___.
3. Some are built by the piling up of ejected fragments around the vent in the shape of a cone with a central crater.

WATER CYCLE
1. This describes the continuous movement of water on, above, and below the surface of the Earth.
2. It is also called the hydrologic cycle.
3. The sun is an important part of this because its heat evaporates water from the oceans into the atmosphere to form clouds.

WEATHERING
1. This is the disintegration of rocks into small soil particles through the physical and chemical action of atmospheric agents.
2. Agents responsible for this disintegration of rocks include rain, water, frost, wind, changes in temperature, plant and animals.
3. Erosion causes something to move away; ___ causes it to be broken into small pieces.

Earth Science Bingo

Earth Science Bingo

Fault	Aquifer	Canyon	Hydrosphere	Lithosphere
Caverns	Atmosphere	Tides	Metamorphic	Water Cycle
Basalt	Weathering		Mountains	Craters
Sedimentary	Eruptions	Volcanoes	Lava	Frederich Mohs
Plateaus	Geology	Quartz	Tsunamis	Tectonic Plates

Earth Science Bingo

Sedimentary	Basalt	Mantle	Seismic Waves	Hemisphere
Frederich Mohs	Earthquakes	Bedrock	Eruptions	Lava
Crust	Geology		Folds	Volcanoes
Rocks	Planet	Weathering	Properties	Tectonic Plates
Water Cycle	Tides	Properties	Caverns	Tsunamis

Earth Science Bingo

Sedimentary	Volcanoes	Earthquakes	Lava	Basalt
Metamorphic	Atmosphere	Core	Aquifer	Igneous
Eruptions	Tides		Latitude	Caldera
Weathering	Crust	Plateaus	Rocks	Mantle
Tsunamis	Caverns	Quartz	Properties	Hemisphere

Earth Science Bingo: Card No. 3

© Barbara M. Peller

Earth Science Bingo

Weathering	Latitude	Canyon	Caverns	Hemisphere
Magma	Continental Drift	Aquifer	Seismic Waves	Basalt
Mountains	Rocks		Lithosphere	Hydrosphere
Volcanoes	Equator	Tides	Properties	Bedrock
Continent	Water Cycle	Planet	Tsunamis	Craters

Earth Science Bingo

Water Cycle	Lithosphere	Eruptions	Bedrock	Caverns
Magma	Volcanoes	Core	Folds	Atmosphere
Canyon	Craters		Metamorphic	Geysers
Tectonic Plates	Hemisphere	Fault	Properties	Erosion
Earthquakes	Quartz	Basalt	Weathering	Mountains

Earth Science Bingo

Caldera	Lava	Mantle	Hemisphere	Craters
Latitude	Eruptions	Erosion	Aquifer	Basalt
Seismic Waves	Continent		Continental Drift	Folds
Properties	Plateaus	Quartz	Planet	Canyon
Frederich Mohs	Bedrock	Fault	Mountains	Equator

Earth Science Bingo

Fault	Lava	Geysers	Metamorphic	Earthquakes
Frederich Mohs	Hemisphere	Geology	Atmosphere	Magma
Mantle	Hydrosphere		Folds	Continental Drift
Weathering	Rocks	Core	Sedimentary	Crust
Properties	Caverns	Quartz	Planet	Caldera

Earth Science Bingo: Card No. 7

Earth Science Bingo

Mountains	Latitude	Fumaroles	Lava	Continental Drift
Magma	Canyon	Seismic Waves	Craters	Bedrock
Equator	Ring of Fire		Hemisphere	Lithosphere
Tsunamis	Weathering	Sedimentary	Continent	Rocks
Tides	Quartz	Planet	Eruptions	Frederich Mohs

Earth Science Bingo: Card No. 8

Earth Science Bingo

Folds	Earthquakes	Geology	Equator	Caverns
Continent	Hemisphere	Mountains	Eruptions	Latitude
Igneous	Fault		Atmosphere	Fumaroles
Erosion	Tectonic Plates	Plateaus	Metamorphic	Geysers
Rocks	Properties	Core	Sedimentary	Lithosphere

Earth Science Bingo

Sedimentary	Lava	Continental Drift	Seismic Waves	Equator
Craters	Bedrock	Aquifer	Atmosphere	Hemisphere
Ring of Fire	Latitude		Hydrosphere	Crust
Plateaus	Tectonic Plates	Erosion	Quartz	Igneous
Core	Frederich Mohs	Mantle	Water Cycle	Mountains

Earth Science Bingo

Caldera	Latitude	Eruptions	Erosion	Frederich Mohs
Fumaroles	Igneous	Metamorphic	Folds	Aquifer
Magma	Hemisphere		Mantle	Geology
Core	Basalt	Properties	Caverns	Sedimentary
Continent	Quartz	Fault	Planet	Earthquakes

Earth Science Bingo: Card No. 11

Earth Science Bingo

Earthquakes	Lithosphere	Igneous	Lava	Folds
Geology	Frederich Mohs	Canyon	Planet	Atmosphere
Fault	Geysers		Craters	Seismic Waves
Quartz	Rocks	Hemisphere	Sedimentary	Magma
Latitude	Fumaroles	Ring of Fire	Continent	Bedrock

Earth Science Bingo: Card No. 12

Earth Science Bingo

Erosion	Lithosphere	Caldera	Igneous	Craters
Canyon	Fumaroles	Hemisphere	Folds	Crust
Latitude	Earthquakes		Geology	Geysers
Mountains	Properties	Continental Drift	Ring of Fire	Sedimentary
Quartz	Tectonic Plates	Planet	Fault	Metamorphic

Earth Science Bingo: Card No. 13

Earth Science Bingo

Caverns	Hemisphere	Eruptions	Folds	Continent
Bedrock	Fault	Igneous	Atmosphere	Latitude
Erosion	Hydrosphere		Mantle	Core
Tectonic Plates	Properties	Ring of Fire	Continental Drift	Caldera
Quartz	Seismic Waves	Crust	Frederich Mohs	Mountains

Earth Science Bingo

Metamorphic	Folds	Eruptions	Earthquakes	Lava
Caldera	Mantle	Aquifer	Canyon	Continent
Craters	Fault		Basalt	Latitude
Quartz	Igneous	Fumaroles	Properties	Erosion
Frederich Mohs	Rocks	Planet	Equator	Geology

Earth Science Bingo

Continental Drift	Igneous	Fumaroles	Equator	Seafloor Spreading
Seismic Waves	Crust	Geysers	Magma	Hydrosphere
Erosion	Lithosphere		Craters	Geology
Weathering	Bedrock	Quartz	Minerals	Sedimentary
Continent	Tephra	Planet	Rocks	Latitude

Earth Science Bingo: Card No. 16

Earth Science Bingo

Core	Minerals	Glacier	Igneous	Caverns
Metamorphic	Continent	Properties	Hydrosphere	Geysers
Folds	Mountains		Tephra	Fumaroles
Tectonic Plates	Frederich Mohs	Sedimentary	Eruptions	Crust
Plateaus	Erosion	Earthquakes	Latitude	Lithosphere

Earth Science Bingo: Card No. 17

Earth Science Bingo

Equator	Ring of Fire	Bedrock	Erosion	Seismic Waves
Latitude	Core	Plateaus	Craters	Continent
Folds	Crust		Glacier	Canyon
Tectonic Plates	Aquifer	Properties	Sedimentary	Mantle
Tephra	Igneous	Eruptions	Minerals	Caldera

Earth Science Bingo: Card No. 18

Earth Science Bingo

Craters	Caldera	Igneous	Fumaroles	Craters
Metamorphic	Latitude	Lava	Earthquakes	Hydrosphere
Minerals	Caverns		Atmosphere	Basalt
Mantle	Tephra	Plateaus	Rocks	Glacier
Canyon	Seafloor Spreading	Frederich Mohs	Mountains	Planet

Earth Science Bingo: Card No. 19

Earth Science Bingo

Ring of Fire	Minerals	Lava	Igneous	Planet
Bedrock	Geology	Magma	Plateaus	Seismic Waves
Lithosphere	Geysers		Weathering	Aquifer
Water Cycle	Tides	Tsunamis	Rocks	Tephra
Volcanoes	Mountains	Seafloor Spreading	Sedimentary	Glacier

Earth Science Bingo: Card No. 20

© Barbara M. Peller

Earth Science Bingo

Metamorphic	Caldera	Magma	Igneous	Water Cycle
Lithosphere	Glacier	Continental Drift	Fumaroles	Fault
Crust	Frederich Mohs		Minerals	Eruptions
Plateaus	Earthquakes	Tephra	Tectonic Plates	Mountains
Weathering	Seafloor Spreading	Planet	Core	Rocks

Earth Science Bingo

Equator	Mantle	Glacier	Canyon	Erosion
Seismic Waves	Lava	Basalt	Fumaroles	Atmosphere
Bedrock	Hydrosphere		Fault	Geysers
Tephra	Tectonic Plates	Rocks	Aquifer	Caverns
Seafloor Spreading	Core	Minerals	Crust	Magma

Earth Science Bingo: Card No. 22

© Barbara M. Peller

Earth Science Bingo

Continental Drift	Minerals	Earthquakes	Canyon	Planet
Caldera	Ring of Fire	Frederich Mohs	Metamorphic	Aquifer
Mantle	Erosion		Tsunamis	Fault
Crust	Seafloor Spreading	Tephra	Core	Rocks
Water Cycle	Tides	Mountains	Plateaus	Glacier

Earth Science Bingo

Continental Drift	Ring of Fire	Caverns	Minerals	Fumaroles
Glacier	Planet	Magma	Seismic Waves	Fault
Geysers	Equator		Erosion	Crust
Water Cycle	Tsunamis	Tephra	Core	Lithosphere
Volcanoes	Weathering	Seafloor Spreading	Lava	Tides

Earth Science Bingo: Card No. 24

Earth Science Bingo

Weathering	Magma	Minerals	Eruptions	Glacier
Aquifer	Tectonic Plates	Metamorphic	Continental Drift	Atmosphere
Lithosphere	Fumaroles		Tsunamis	Tephra
Basalt	Water Cycle	Tides	Seafloor Spreading	Hydrosphere
Planet	Caverns	Continent	Bedrock	Volcanoes

Earth Science Bingo

Glacier	Minerals	Mantle	Seismic Waves	Equator
Plateaus	Lava	Fumaroles	Ring of Fire	Continental Drift
Tectonic Plates	Tsunamis		Hydrosphere	Weathering
Core	Canyon	Water Cycle	Seafloor Spreading	Tephra
Geysers	Continent	Eruptions	Tides	Volcanoes

Earth Science Bingo

Mantle	Bedrock	Minerals	Ring of Fire	Geology
Water Cycle	Tsunamis	Metamorphic	Tephra	Atmosphere
Quartz	Tides		Seafloor Spreading	Weathering
Equator	Caldera	Magma	Volcanoes	Aquifer
Continent	Hydrosphere	Glacier	Basalt	Geysers

Earth Science Bingo: Card No. 27

Earth Science Bingo

Craters	Ring of Fire	Basalt	Minerals	Continental Drift
Geology	Glacier	Tsunamis	Seismic Waves	Hydrosphere
Tides	Crust		Geysers	Plateaus
Sedimentary	Equator	Frederich Mohs	Seafloor Spreading	Tephra
Canyon	Folds	Continent	Volcanoes	Water Cycle

Earth Science Bingo

Glacier	Ring of Fire	Equator	Metamorphic	Folds
Tectonic Plates	Plateaus	Magma	Geysers	Basalt
Lithosphere	Tsunamis		Atmosphere	Minerals
Geology	Water Cycle	Hemisphere	Seafloor Spreading	Tephra
Continental Drift	Fumaroles	Volcanoes	Caldera	Tides

Earth Science Bingo: Card No. 29

Earth Science Bingo

Caverns	Minerals	Seismic Waves	Folds	Tephra
Aquifer	Ring of Fire	Mantle	Hydrosphere	Atmosphere
Tectonic Plates	Erosion		Geysers	Magma
Volcanoes	Caldera	Canyon	Seafloor Spreading	Tsunamis
Water Cycle	Earthquakes	Tides	Glacier	Basalt

Earth Science Bingo: Card No. 30

www.ingramcontent.com/pod-product-compliance
Lightning Source LLC
Chambersburg PA
CBHW051418200326
41520CB00023B/7285